Mechanical keyboards from switches to keycaps

CHRISTOPHER DOCK

ISBN: 978-1-952930-03-4

North American Offices
Christopher Dock
3533 Rum River Drive
Anoka, MN 55303
USA

European Offices
Christopher Dock
Ludwig-Richter-Straße 15
60433 Frankfurt
Germany

Limits of Liability and Disclaimer of warranty

We work hard to provide the most accurate research on mechanical keyboards and associated technology. All information provided in Mechanical keyboards from switches to keycaps

The information provided in this book is about the options available for on building and purchasing mechanical keyboards or parts, and should not be considered as advice or recommendations for selecting any individual parts or completed keyboards.

CONTENTS

DEDICATION

For my family whose patience and
encouragement
made this book possible.

INTRODUCTION

Perhaps the most understated peripheral of a computer is the keyboard. Every computer purchased from a big computer manufacturer tends to bundle one with the sale and in many cases free of charge. It is nice receiving a free gift but like most things you get what you pay for.

Figure 1: Generic keyboard

It is also possible to purchase one of these keyboards new from your favorite online website for between 10 and 15 dollars. These keyboards are suitable for their task but they tend to be built with the cheapest quality materials possible in order to meet this retail price point.

These modern miracles of engineering print the electronic circuits onto a few pieces of plastic and have a small rubber dome, Figure 2, which both prevents the key from registering until it has been pressed but also to provide some feedback to the user.

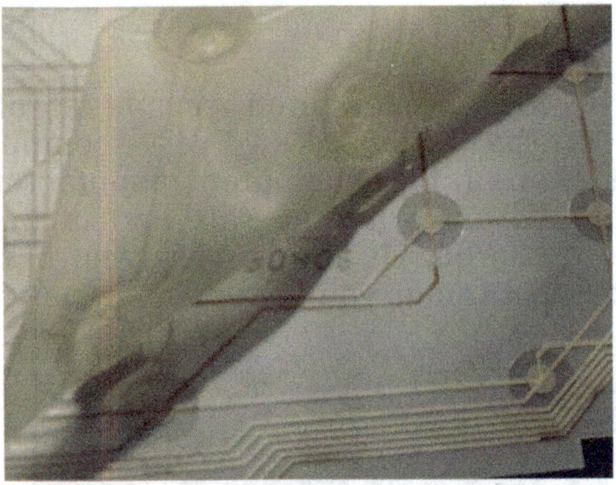

Figure 2: Rubber dome keyboard electronics

It wasn't always this way. Back when personal computers were a new commodity all keyboards[1] were a mechanical keyboard. The keys on this keyboard were a mechanical affair where

1 All IBM personal computer keyboards

pressing down the key caused a spring to bend (buckle) until the bent spring caused the hammer to pivot forward completing the circuit and registering a key press for that key.

One of the most important features of this early mechanical keyboard was the tactile feedback this key provided. The typist would be able to feel a small bump as the key-press was registered. This feedback was invaluable to touch typists as they did not need to continue to press the key any further as that would cause it to bottom out but not actually generate any additional information. This allows the typist to type faster as they only needed to press the key the minimum amount required. In addition to this tactile feedback there was also an audible click that occurred when pressing the key to the activation point.

Many of these original mechanical keyboards still work today and are quite sought after for reasons of nostalgia but also because they do have a very distinctive feel that is enjoyable to use.

These mechanical works of art cost so much to create becaues of the all of the moving parts involved - quite literally. They were probably costly to replace due to all the hand assembly required to create them.

Yet these keyboards were built with a very high quality and because of this many still can still be found in working order today. These IBM model M keyboards even when not in fully working order can still command a few hundred dollars in certain circles.

IBM's keyboards were both complicated to build as well as expensive but IBM did not have a monopoly on engineering expertise. The Cherry Corporation created and patented the iconic Cherry MX switch, Figure 3, as an alternative part in mechanical keyboards.

Figure 3: Cherry MX switch

This humble switch help launch the market segment of third party mechanical keyboards.

1 THE BASICS

It is easy to overlook all of the different parts that make up a mechanical keyboard. The most basic list of components for a mechanical keyboard is as follows.

- Case
- Base plate
- Circuit board
- Stabilizers
- Switches
- Keycaps

A lot of these parts seem pretty obvious such as the keyboard case or the keycaps but what are stabilizers and what is the base plate?

Stabilizers, Figure 4, help long keys to work as they intuitively should. It would not be a very nice typing experience to press on the left side of the spacebar and only that side goes down. It would also feel odd if the shift or backspace keys seemed to rock if you did not press it exactly in the center of the key. The stabilizers are designed to depress the entire key uniformly no matter where on the key you press down.

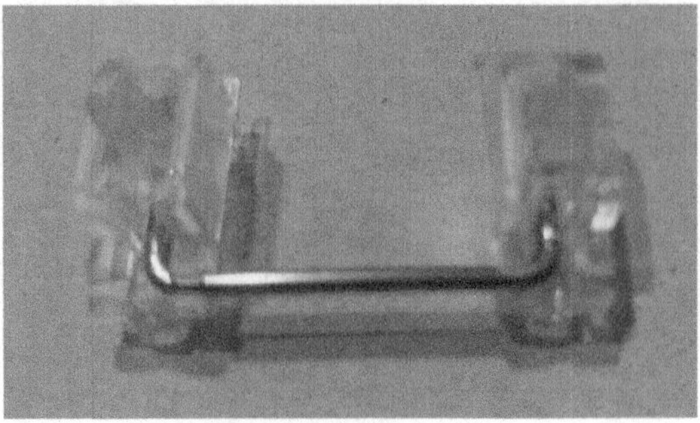

Figure 4: Stabilizer

The base plate, Figure 5, is used to help align all of the key switches with the circuit board but it does more than just that. It also provides stability and structure to the entire keyboard. The cheapest and flimsiest keyboard case will not droop when you pick it up by one end but the addition of the base plate also guarantees the keyboard will feel solid in your hand, prevent flex when typing and have a certain amount of weight when holding it.

Figure 5: Base plate

A keyboard case is no different than most consumer commodities, it holds all of the parts for the device. For a keyboard the case is composed of two or more parts and the base plate forms the greater part of the top of the keyboard.

Keyboard cases are typically made of plastic but some of the higher end cases are made from a CNC milled piece of aluminum. Some high end cases even have a slot where a denser metal bar can be screwed in to add more stability and weight to the case.

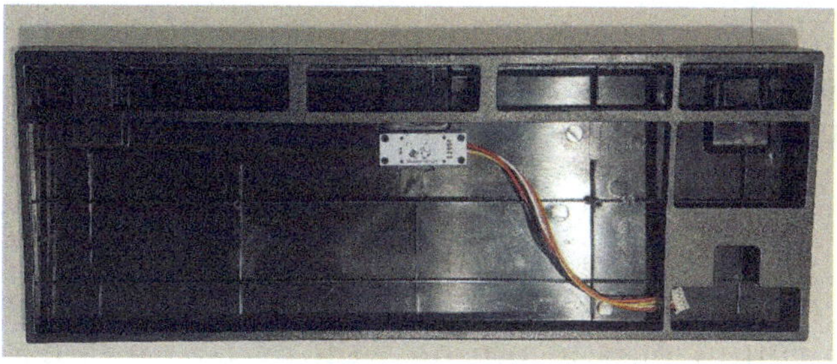

Figure 6: Keyboard case

Keycaps are probably the most well known part of a keyboard, mechanical or not. They have some connector on the underside that will allow it to be attached to the top of a key switch. There is much more behind the humble keycap but that will be covered later.

Figure 7: Keycaps

The second most visible item on some keyboards is the key switch which both holds the keycap and allows the users key press to be captured by the keyboard itself.

All of these mechanical keyboard key switches look superficially alike. On the bottom of the switch, the base housing, is usually one very large round plastic pin as well as two metal pins which is used for the electrical connection with the circuit board. Sometimes there are two additional plastic pins on the bottom that are used for keeping the orientation of the switch parallel with all other switches.

The upper housing can be made of different materials or can be a slightly different shape. Sometimes it is even designed with a spot for a led.

Figure 8: Key switches

The circuit board of the keyboard is just like the motherboard for your computer except it is much simpler. It only needs to test for switches being pressed and to communicate that information with the connected computer. Sometimes the circuit board has additional duties such as controlling status leds or led lighting effects.

I have not found any standards for the actual sizes of the keyboard circuit boards. The circuit board is designed for its case and the mounting holes must match to the case itself.

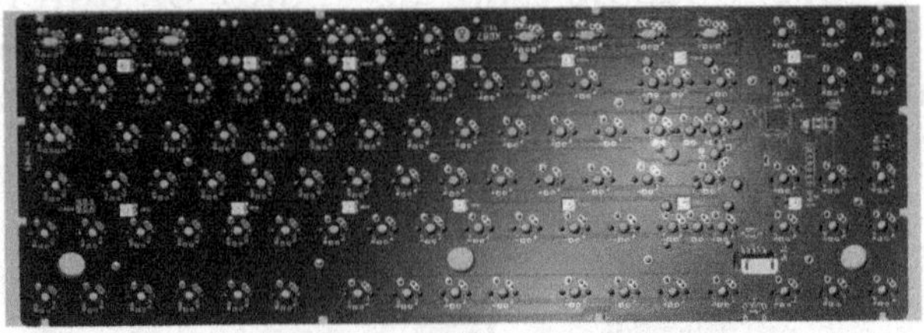

Figure 9: Circuit board

Keyboard layouts

Despite the fact that these above described parts are used in all mechanical keyboards the end results do not look the same. It might seem obvious that each manufacturer, Cherry, Razer, Roccat, Logitech and many many more choose to style the case differently so they look subtlety different from their competitors.

Despite keyboard character layouts being rather standard there are still quite a number physically different layouts available. The keyboard layout that is the most familiar is the full 104/105 key keyboard

Figure 10: ISO 105 key keyboard

The keyboards have one or two basic keyboard layouts, either ANSI or ISO. The ANSI keyboard is found in the United States while the ISO keyboards are found in the rest of the world. It is easy to tell the difference between the two by looking at the return key. The return key on an ANSI keyboard is essentially a simple rectangle while on a ISO keyboard it looks more like a big fat upside down letter L.

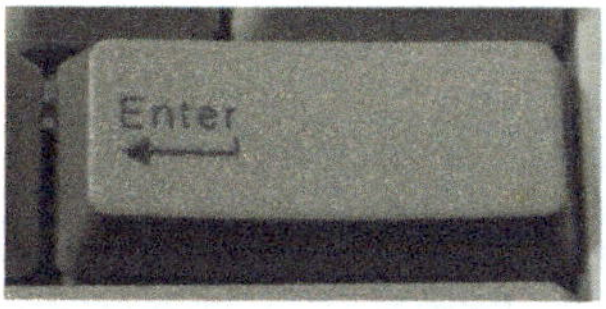

Figure 12: ANSI Enter key

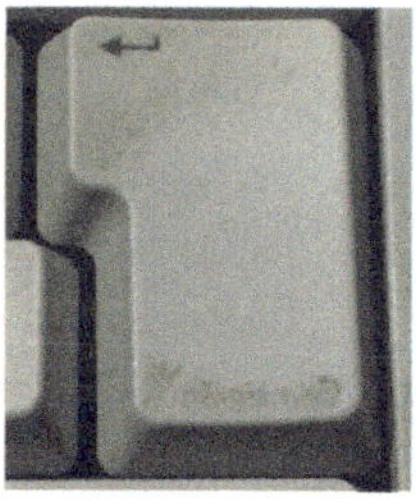

Figure 11: ISO Enter key

However, the real keyboard differences are whether the keyboard is a 104 or a 105 key keyboard. The 104 key model is the standard

ANSI Qwerty keyboard. It is actually fairly simple as you only need the shift, alt or control key to generate any of the keys that are printed on the keycaps.

This is a different story for the various "foreign" keyboards whether they be a Spanish Qwerty keyboard or a German Qwerz keyboard. Each has additional regional characters or symbols when compared to the standard American Qwerty keyboard but still different from each other.

In order to use some of these special characters you will need to use the 'altgr' key. The altgr key, in much the same way as the shift or control key, makes it possible to select any of the additional characters. It is interesting to note that the physical hardware for 105 key keyboards is essentially identical to 104 key keyboards.

The 104 and 105 key layout has all the keys. It has a numeric keypad, arrow keys, function keys as well as other movement keys. This rather generic keyboard has everything with the trade off that it takes up a considerable amount of space on your desk.

For those who do not need a dedicated numeric keypad on their there is a more compact keyboard available.

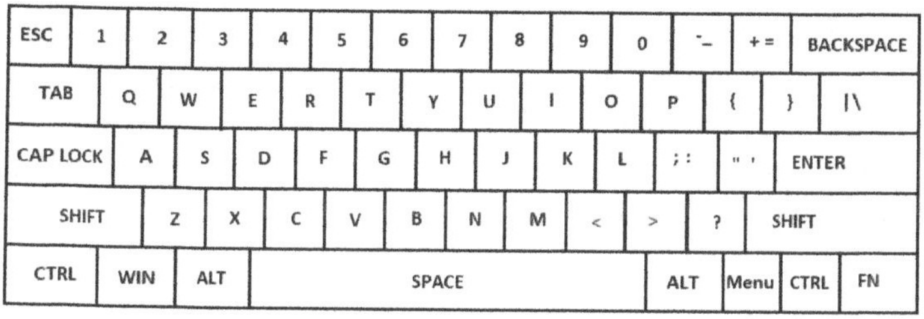

Figure 16: Example of a 60% keyboard

This keyboard is actually much cheaper to build as the number of switches is less and it is easier to find both circuit boards and cases for this size. Partly because it is smaller it is easier for home electronic hobbyists to create this type of keyboard without too much difficulties.

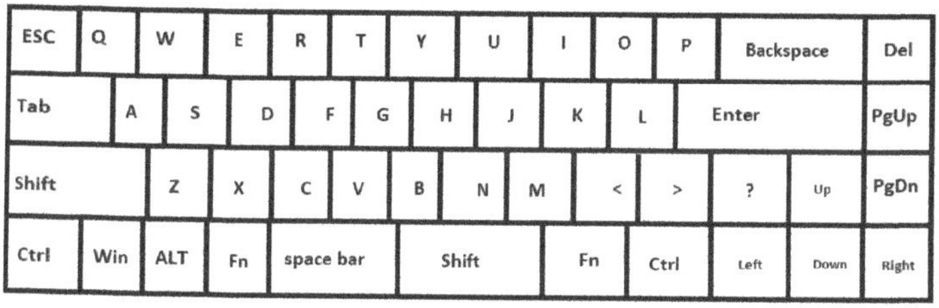

Figure 17: Example of a 40% keyboard

The smallest keyboard is the 40% keyboard. It is pretty much a unicorn when it comes to finding one in the wild. I have actually seen variations of most of these keyboards but never actually seen a 40%.

In theory these optical key switches should be slightly faster. Not so much because of the light but because there is no physical contact there is no need for the key debounce delay.

One thing not described so far is from where this light is generated. This can vary depending on the type of the optical switch but it seems likely that this light would be coming from the circuit board up into the key switch.

What is important is that each optical key switch manufacturer has implemented their switch differently. So if a optical key switch needs to be replaced then it must be replaced by the same brand key switch. Yet, this also means we cannot use optical key switches as drop-in replacements for any mechanical keyboard.

3 STABILIZERS

The stabilizers are for the most part standard but they may not actually be identical across all manufacturers. This is especially important when the keyboard in question is a "store bought" brand.

Large keyboard manufacturers can design their own key switches and stabilizers instead of using more standard components and there is no guarantee that these stabilizers can be replaced with after market stabilizers.

In addition to this is the fact that there are two basic classes of stabilizers. One type of stabilizer clips directly onto the circuit board while the other clips onto the base plate. As these are not compatible with each other you need to pay special attention when ordering a set of stabilizers separately from your circuit board and base plate.

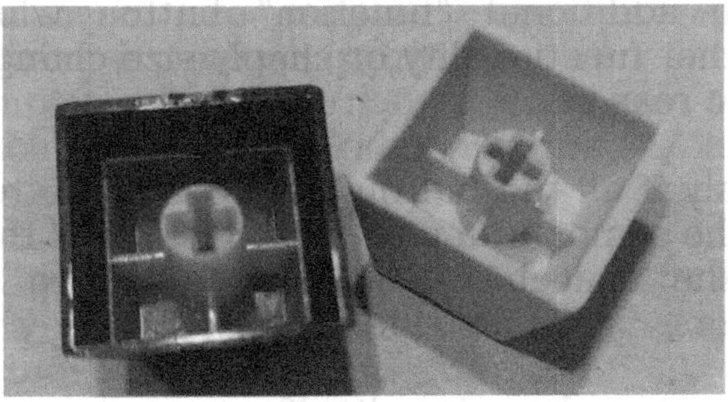

Figure 31: Normal key (left) and after market key (right)

ANSI is king

In chapter 1 there were a lot of different keyboard layouts. These layouts ranged from full 104 key keyboards down to much more compact 40% keyboards.

These keyboard layouts are only a subset of total number of keyboards because many different languages have special non-ANSI letters and their keyboards reflect that. Thus all of those keyboard layouts will have alternatives that support other languages.

It is odd that with so many people whose native language is not English it is difficult to find more than a few sets of keycaps that support these other countries (ie Germany, France, United Kingdom,...)

I have managed to find a few on the Amazon marketplace but they were actually sold as UK keycaps that supported German ISO layout.

Figure 32: YMDK Laser Engraved UK Keycap for Mechanical MX Keyboard

The UK keycaps, Figure 32, actually would have been fine for a UK keyboard but because it was also sold as an ISO German it contained a few extra keys as well, Figure 33.

Figure 33: Additional keys for German ISO

Not an optical illusion

If it were not enough to know that different keycaps, such as alt or control, might have a different size on different keyboards you might be disillusioned to learn that the keycaps may also have different profiles.

Keyboards is not only a collection of keys in different combinations they have their own profile. They can be low profile or high profile. The low profile keyboards are essentially flat like the chiclet style keycap, Figure 34.

Figure 34: Chiclet style profile

Figure 35: Flat or staircase profile

Another low profile keycap is the flat style which can be seen in Figure 35. This style is sometimes called a staircase if the keyboard has a big slant.

Figure 36: curved profile

Figure 37: OEM profile

The high profile keyboard can have a contoured or curved keyboard similar to Figure 37. This does not exclude the possibility that a low profile can have a curved profile but it is less common.

Of all of the keycaps that I have seen for sale either have been marked as having an OEM profile or does not mention the profile at all.

The OEM profile has a different shape for each row. This is what gives the concave appearance. When the key are not in their proper row you will see a bunch of keys that do not seem to match up, Figure 38.

Figure 38: Random keys from an OEM profile set

5 BASE PLATE, PCB AND CASE

Creating a your own keyboard case, base plate and circuit board is pretty much beyond the ability of the most dedicated mechanical keyboard do-it-yourselfers. The tolerance between the different keys on the base plate is obviously very tight. The cutout for Cherry switches on the base plate, Figure 39, show the measurements for a Cherry MX switch.

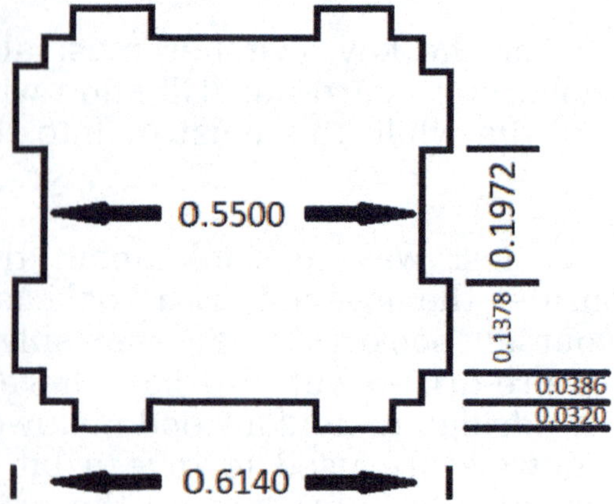

Figure 39: Cherry MX cutout dimensions in inches

The base plate is optional but it is great for both ensuring that all of the switches are properly aligned with the circuit board and to give some extra strength to the keyboard. The type of material that is used for the base plate can affect the sound when using the keyboard. Base plates made out of stainless steel or brass cause higher pitch sounds while base plates made out of lower density materials have lower pitch sounds.

Up until now most of the focus on the switches has been is the stem compatible and what attributes do each of the types have when using the switch. On the other side of the switch is just as important as it contains the electrical contacts as well as stabilization pins. The circuit board needs to have cutouts for the stabilization pins as

well as the connection points for the switch.

These cutouts for the key switches must also align up with the circuit board but still align with the base plate all the while this must fit into the case.

Perhaps the easiest way to think about this would be to consider the assembly of a bookcase. Most build it yourself bookcases have not only a number of holes pre-drilled but also have have a few holes that are drilled to hold a wooden dowel. These wooden dowels are used to give a bit of extra support but also to help align up the parts as you assemble the bookcase.

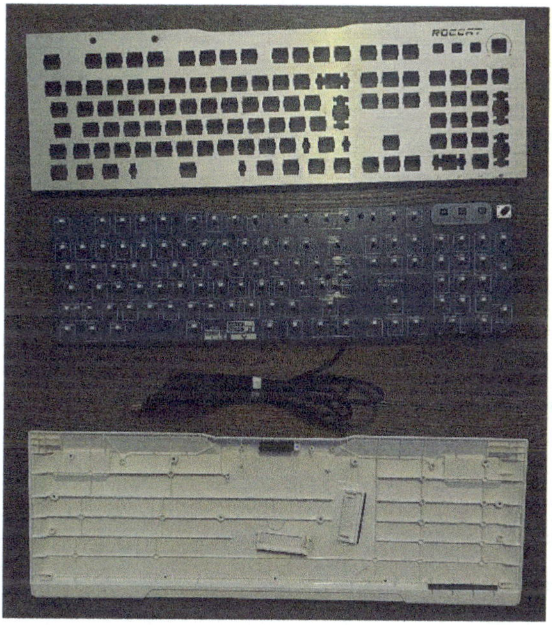
ROCCAT

6 TOOLS AND CONSUMABLES

Unlike some other hobbies such as electronics or embedded computing, mechanical keyboards does not require much in the way of tools.

- Keycap pullers
- Key switch pullers
- Key switch openers

The majority of the keyboard is made up of switches and keycaps so it shouldn't be a surprise that the tools are for precisely these three tasks. The most basic tool is the keycap puller. It might be possible to pull the keycap off purely with your fingers you do risk damaging the switch.

Figure 40: Razer switches, one with broken stem

A keycap puller is a faster and safer method of removing the keycaps. There may a few other variations but these, Figure 41, are pretty common. The keycap puller with the red handle is less damaging to the keycap because the wires slip underneath the keycap and do not scratch the side of the keycap which can happen with the smaller white keycap puller.

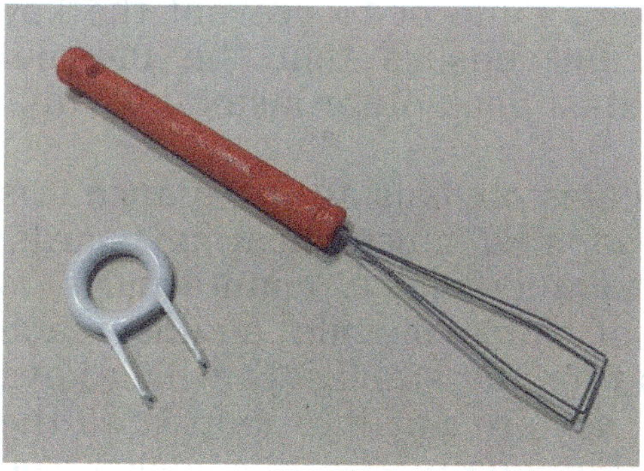

Figure 41: A couple of keycap pullers

It is possible to open up a Cherry MX switch with nothing more than a flat-head screwdriver but a key switch opener, Figure 42, is much faster and there is less chance of damaging the switch.

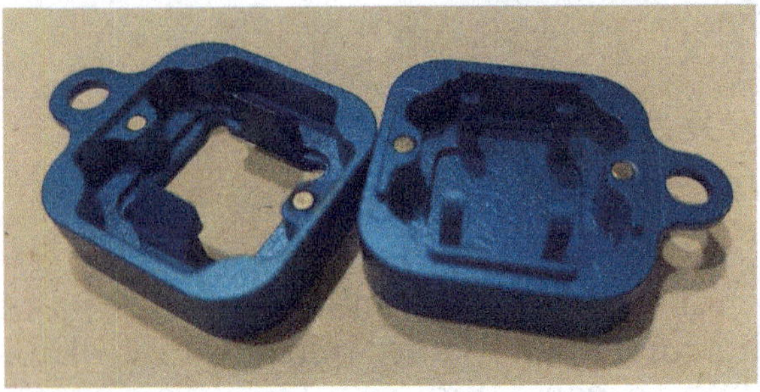

Figure 42: Key switch opener

The key switch opener is only necessary if you plan on lubing the switches. Just like any other mechanical object, friction can take its toll. In the case of mechanical key switches the grease is not to give us a longer life or to prevent the key from locking up, but ensure that the internal workings glide against each other instead of rub.

Although this seems obvious at first glance it is important remember that mechanical keyboards, more than their non-mechanical counterparts, is a big collection of parts moving against each other. This is true for the stabilizers which have small metal rods that are put into motion as the key goes down and up again. But this is also true for the stems of the mechanical keyboard which is sliding up and down within the housing for every keypress.

Lubing the key switch is one of the primary modifications made to mechanical keyboards but this can only be done prior to soldering the switch into the circuit board. The process is pretty simple, you disassemble the switch and put some grease into the switch with a small paint brush. In addition to this you carefully grease up the spring. Lubing each spring independently would take forever but this can be done by putting all of the springs into a small plastic bag with a small amount of grease and then shaking the bag.

Lubing the switches is a rather time consuming task and because each switch is comprised of a few pieces. If you are planning on doing this you may wish to invest in a lube station. This is just a board with cutouts to hold all of the different parts. A lube station has a spot for each part of the switch, the stem, the spring and the base. It may also have a depression where you can put the grease can.

Silky smooth key presses is an alluring goal but there is one note of caution. Indeed the feel of the switch, especially linear switches, will change usually for the better. But this change also includes a slight change in the noise profile. Quieter might be desirable in an office setting but some key switches are not only tactile but actually designed to be especially clicky when depressed. The tactile feel will remain but the sound or tactile or clicky key switches will be altered forever. This change to the key switch is a one way trip. Almost every other modification can be reversed except for lubing a key switch.

- remove dampening materials
- use old keycaps
- remove o-rings
- use previous base plate
- add or remove film stabilizers

One additional modification that can be done, probably after lubing the switches would be to use film stabilizers on the key switch. This film is a very thin plastic or rubber that is added between the top and bottom housing of the switch. The film is very thin can be quite fiddly to use, so you may wish to have a few extra in case problems occur while trying to add them to the switch.

A much easier modification that can be done is to add o-rings to the keycaps, Figure 43.

Figure 43: O-ring modification

The sound can be reduced by using linear key switches or by adding o-rings. Adding an o-ring is not at all the same quality enhancement as lubing the switch. O-rings can be added if you find the keyboard a bit too loud. The sound can be generated by the switch or perhaps by pressing the keys so hard that they actually hit the base plate.

One final modification is to add some type of dampening material into the case. It doesn't matter if the keyboard is a high end from Razer or Logitech or a entry level the case might have a bit of a hollow sound when typing on it. This sound can be eliminated by purchasing some sort of dampening material, ie felt, and adding it into the case underneath the circuit board.

This is also a easy modification that can be performed on any keyboard without removing all of the switches or taking apart a hundred switches.

Up until now there has been no real discussion of soldering anything. However, all of those switches need to be attached to the circuit board. Soldering through hole parts is actually really easy but to do so you need to have a soldering iron and solder. More items may be necessary if it is possible, and it is possible, that mistakes can be made that may need to be fixed.

Soldering is the process of joining different types of metal together with melted solder. Solder is a metal alloy made of tin and lead but sometimes includes traces of bismuth, copper, indium, nickel or silver. These trace metals give different attributes to the solder as it heats up or cools down.

A soldering iron can be a cheap and easily

obtained tool which can generate up to 400C on a metal tip which is pressed against the metal to be joined as the solder is applied.

Soldering a switch to the circuit board makes for a very stable bond. Once this is done the switch will be rock solid without any wobble. Of course if you solder the switch so it is not level or square with the rest of the keyboard then it will be that way forever as well.

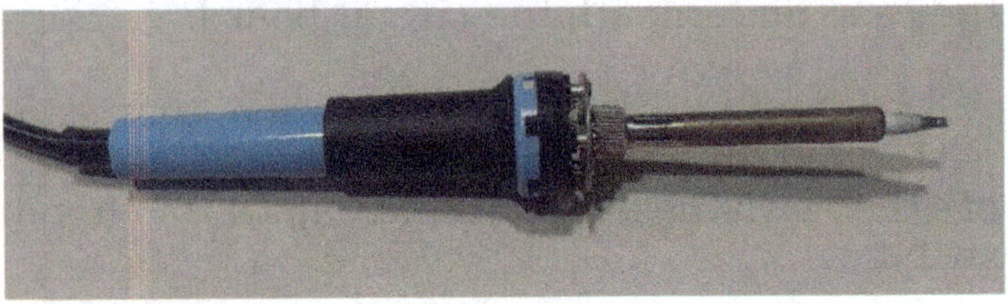

Figure 44: Soldering iron

Not everyone owns a soldering iron, Figure 44, and this can be a barrier to entry when it comes to creating your own mechanical keyboard. Even for those who do own a soldering iron the permanence of soldering in a key switch is a bit of a drawback. Key switches are expensive but if you don't like the feel of the A S D F keys or need a lighter touch for some keys you might be facing hours of desoldering and resoldering to make those changes

It is outside of the scope of this book to cover all of the tools and consumables typically associated with soldering but if you are interested there is a very friendly document[4] on the internet, Soldering is easy, which describes how to do through hole soldering.

Figure 45: Soldering guide

Some years back someone had a clever idea to deal with this problem. Much like it is possible to have a hot swappable disk drive in your server it is now possible to have hot swappable key switches in mechanical keyboards. The key switches plug into special sockets which are soldered to the board. Thus at a later point you decide to change some of the switches you only need to remove the keycaps and carefully pull the

4 http://www.netzmafia.de/skripten/hardware/FullSolderComic_EN.pdf

switches out.

This is a fabulous improvement but it then adds additional costs to the keyboard itself. This might add an additional 15 - 20 dollars for these hot swappable sockets to the cost of the keyboard.

7 LESSONS LEARNED AND EXAMPLES

Keyboard #1
I purchased a Logitech G710+ mechanical keyboard with the intention of personalizing it at a later date. When that time came I realized that the bottom row of modifier keys were an unusual sizes when compared with the keycaps that I found online at that time.

Lesson learned
It is important to find a mechanical keyboard that you like yet if you are considering switching keycaps in the future you should check to see if the keys are standard sizes before purchasing either the keycaps or the keyboard. This should apply for all keys but the main differences tend to be on the modifier row keys (alt, control, window, etc).

Keyboard #2
A few ago I happened to see a few broken keyboards in the bargain bin and I thought purchasing them would be a bargain. The first keyboard was missing the keycaps and it had a single bad switch. Replacing the key switch only took a few minutes but I was surprised to later find out that the stabilizers on the keyboard, Razer Blackwidow Elite, wouldn't support the keycaps that I had purchased.

It seems obvious if the keycaps don't fit on the stabilizers then you just need a new set of stabilizers. Yet even this idea wasn't quite correct. I did purchase a set of plate mounted Cherry MX stabilizers; when they showed up I couldn't actually use them. It was an embarrassing oversight but the stabilizers for the keyboard were custom to that Razer keyboard and a generic set of Cherry MX stabilizers were too wide and would not fit. The keyboard is actually in perfect working order but it is not possible to find any stabilizers that fit. Thus, the Razer Blackwidow does support the Cherry stemmed keycap set that I own.

It is a pretty tough dilemma. The keyboard originally had a UK set of keycaps which I thought I could replace with an after market set of German keycaps. This is not possible because the stabilizers are not Cherry MX compatible. It does not seem possible to purchase Cherry MX compatible stabilizers for this keyboard, but the Razer website does not sell German keycaps, only for the US and UK layouts.

The Razer Blackwidow Elite that I have is a 105 key keyboard. The US ANSI keyboard is a 104 key keyboard so the keyboard is not really suitable for an after market ANSI keycap set.

Lesson learned
Big brand keyboards tend to offer high quality keyboards with lots of functionality. Yet the manufacturers need to stand out from other keyboards as well as compete on cost and some of the ways this has been done is with custom key switches, non-standard key switch stems, or non-standard case/circuit boards sizes.

Keyboard #3
Another keyboard from the bargain bin that I had hoped to re-use did not turn out that way. The keyboard was the Roccat Vulcan 120. This keyboard had no mechanical / electrical defects but was missing a few keycaps.

This is an absolutely beautiful RGB keyboard but it uses half height keycaps. This is partly because of the shape of the Roccat Titan key switches. It is difficult to use normal Cherry MX keycaps, I cannot say this is true for all but it was true for the set of keycaps that I did purchase.

Because of my experience with Keyboard #2 it was possible to see that the base plate is not sized for standard Cherry MX stabilizers. This is a great keyboard but hard to modify with some standard parts.

Lesson learned
Name brand keyboards are amazing but they

tend to be built with some proprietary parts and because of this it can be difficult to modify if need be.

Keyboard #4
I wanted to build a keyboard from scratch but I tried to cut some corners. I did find a company on Alibaba Express that was selling all of the parts for a mechanical keyboard.

- Circuit board
- Case
- Base plate
- Stabilizers

Going cheap is not a good idea. The quality of the items I received was not quite up to the standard I was expecting. The set of stabilizers was incomplete and one of them was misshapen. I have been having problems getting the base plate, circuit board both into the case with the stabilizers.

Lesson learned
Purchasing on Alibaba Express is not really harder to use than Amazon but only once you find the items. This might be difficult due to the size of the site. Purchasing items from the other side of the world where they don't natively speak your language can have its challenges. There was initially problems with entering addresses, or specifically addresses that do not match United States format.

It does not matter where you purchase your parts, custom building mechanical keyboards is a rather expensive task. In the end, you do get what you pay for.

Keyboard #5

I am a fan of the tenkeyless keyboards so when I saw one in the bargain bin, even with no warranty, I grabbed it. I was a bit surprised that all keys worked except for the letter R key.

Lesson learned

A lot of mechanical keyboards make a big deal about how their key switches will last tens of millions of key presses.

I believe that these key switches will indeed last this long but I have found a few keyboards in the bargain bin where a single key did not work. I simply desoldered those keys, as well as cleaning the circuit board in that area, and soldered in a new key switch. The keyboard then worked again.

The tens of millions of key presses is great but spilling anything on a mechanical keyboard tends to fall short of the cheaper rubber dome keyboards. Those keyboards have a giant piece of rubber that tends to protect the keyboard from all but the most major spills.

Yet, the ability to replace a single switch is a major advantage.

The perfect switch

It is very easy to test out a keyboard model or two at one of the big box electronic retailers. Don't fall in love with the key switches from that keyboard. There number of different key switches are not quite as plentiful as the snowflakes in South Pole but they are close.

Lesson learned
This book actually has listed a very small subset of Cherry MX as well as the Gateron compatible switches in chapter 2. These are only just a few of the switches that are possible. There are some keyboard enthusiasts who have gathered much more comprehensive lists[5] but they can be difficult to find.

Even shopping on the mighty Amazon websites only a few of the most popular key switches can be purchased at any price. If you fall in love with a very specific model you may need to do a group buy with other enthusiasts or try and find site that does specializes for certain components. This seems obvious but the internet is also full of a lot of people who have had bad experiences when making purchases far off the beaten track.

Features and name brand keyboards
Not all led keyboards are equal and you need to read the description very carefully. The easiest thing to note is either there are no led lights or the leds are a single color (ie red). Some keyboards advertise RGB rainbow colors. Usually this means that there are a few different colors that light up the keyboard in different zones but they remain that one color and do not change.

5 https://thegamingsetup.com/gaming-keyboard/buying-guides/keyboard-switch-chart-table

This is different from the true RGB keyboards which all the keys can take on any color. These true RGB keyboards have different patterns to show off the leds and they are gorgeous, although they tend to be on the expensive end of the spectrum.

It is important not to under estimate the value the big keyboard manufacturers bring to the table. Sometimes the keyboard has a built in USB port or ports. Quite often they support various multimedia keys, launch various windows programs, support WiFi or Bluetooth. Some models even have support for recording macros and replaying them at the press of a key. However, usually this functionality is supported via driver software and not directly by the keyboard. This is great for windows but less impressive for other operating systems such as Linux.

There are quite a few pros and cons of to make or to purchase a mechanical keyboard.

Final takes

Mechanical keyboards have a great feel when typing but there is always the question to build or to buy.

Building a custom keyboard is

... expensive
... often requires soldering
... easy to purchase incompatible parts

I have put together a few possible examples

Example 1 – upgrading an keyboard

Entry level mechanical keyboard	$35
O-rings	$6
Dampening material in case	$5
Keycaps	$25 - $50
Total	$71 - $96

Example 2 – purchasing a kit

Keyboard circuit board	$35
Keyboard case	$15
Base plate	$17
Stabilizers	$18
Key switches	$55
Dampening material in case	$5
Keycaps	$25 - $50
Total	$170-$195

Example 3 – a kit with customization

Keyboard circuit board	$35
Keyboard case	$15
Base plate	$17
Stabilizers	$18
Key switches	$55
Dampening material in case	$5
Lube	$10 - $20
Lube station	$17
Model paint Brush	$5
Key switch opener	$10
Keycap puller	$5
Keycaps	$25 - $50
Total	$217-$252

Example 4 – purchasing a commercial solution

Razer BlackWidow Elite	$139

Example 5 – purchasing a commercial solution

Logitech Mechanical keyboard G915	$175

Without a doubt the rewards for a customized mechanical keyboard are great and if they are treated well they could easily last decades but like many other activities it is expensive.

It has been said before but you get what you pay for.

ABOUT THE AUTHOR

Christopher was born in the United States in the Midwest but apparently became restless and made the move to the Europe. He has spent most of his time working in IT departments for banks and large corporations, usually as a consultant but also as a software developer. In that environment you cannot choose your own equipment and what is more personal to a developer, or a gamer, than a keyboard.

It was his interest in electronics which led him down the path towards building his own mechanical keyboards. Building mechanical keyboards is not overly difficult but there are a lot of little gotcha's that can increase both the time and cost to create your own keyboard. This book is a collection of all the technologies in order to reduce the pitfalls when building your own keyboards.

Made in the USA
Monee, IL
03 November 2026

67380171R00046